ANTARCTICA

The seaward margin of the Sorsdell Glacier near the Vestfold Hills in Princess Elizabeth Land. Rock fragments, ground off the subsurface rocks of the Antarctic continent by moving ice, are revealed in a perfect arch at the coast.

ANTARCTICA

FUTURE OF A FROZEN WILDERNESS

Rob Gell

Houghton Mifflin Australia

To Nicholas, who was a baby when I went there;
to Kimberley, who has learned so quickly to appreciate this Earth —
that they might always have the chance to see it as I did —
and to their mother Jo,
who continues to provide the indulgence, inspiration
and love that I need and cherish.

Houghton Mifflin Australia Pty Ltd
PO Box 289, Ferntree Gully, Victoria 3156
112 Lewis Rd, Knoxfield, Victoria 3180, Australia

First published 1989

National Library of Australia
Cataloguing-in-Publication entry:

Gell, Robert.
Antarctica. Future of a frozen wilderness
ISBN 0 86770 092 0.

1. Ecology — Antarctic regions. 2. Conservation of natural resources — Antarctic regions. 3. Antarctic regions — Description and travel. I. Title.

919.8′904

Edited by Carla Taines
Designed by Derrick I. Stone
Typeset in Palatino by Bookset Pty Ltd
Printed in Hong Kong by South China Printing Co. (1988) Ltd

CONTENTS

The Antarctic continent is only 2700 kilometres south of Australia. However, the territory that Australia claims and Australia's scientific bases lie some distance to the west. Mawson base is south of India and 7000 kilometres from Hobart.

ANTARCTICA: AN INTRODUCTION

Antarctica is the last remaining wilderness continent on earth. It is physically and ecologically unique. It has existed in its present form for at least 13 million years, probably more than twice that. There are no trees, only two flowering plants; no land-based reptiles or mammals; no rainfall and surprisingly little snow. It is a cold, white desert. Yet to be fully explored, it is the only major landmass on earth that has not been taken by conquest, and is the only continent where man is not a native inhabitant. But like so much of the earth, Antarctica is now feeling the impact of human intrusion, and as we approach the end of the twentieth century, man is exerting even greater pressure on the frozen continent. Once the realm of the discoverers and explorers, Antarctica is now being eyed greedily both as a rich mineral resource and as a tourist destination.

Antarctica is magnificent, a vast icy continent guarded by the Southern Ocean with its ferocious storms and sentinel icebergs. It took man almost four centuries to reach Antarctica, and only since substantial protective dwellings have been built have humans been able to establish a toehold on the small ice-free areas on its margins — the same ice-free areas that are used by the seals, penguins and other birds for breeding colonies.

There are now fifty bases, representing twenty Antarctic Treaty nations, established in the Antarctic. They all have an impact on their immediate environment, and many on a wider area. The American base at McMurdo is the largest of all — it accommodates a thousand people, and 4000 tonnes of re-supply equipment each year, but the cost is high. The Southern Ocean in the vicinity of the base is severely polluted, rubbish disposal techniques are appalling, and in some locations the total landscape is being changed. Raised dust from heavy machinery has darkened nearby snow and ice in some areas, changing its albedo, its ability to reflect solar energy, and causing it to melt. The effect of the bulldozer blade is substantial, and the imprint of bulldozer tracks has lasted more than a decade. Similarly, where the pressure of a human footfall leaves a mould in a delicate mossbed, the engraving is likely to remain for tens of years.

EAST OF GREENWICH
EQUATOR
90°
180°
AUSTRALIA
TROPIC OF CAPRICORN
NEW ZEALAND
INDIAN OCEAN
40°
60°
ANTARCTICA
PACIFIC OCEAN
AFRICA
ATLANTIC OCEAN
SOUTH AMERICA
0°
90°
WEST OF GREENWICH

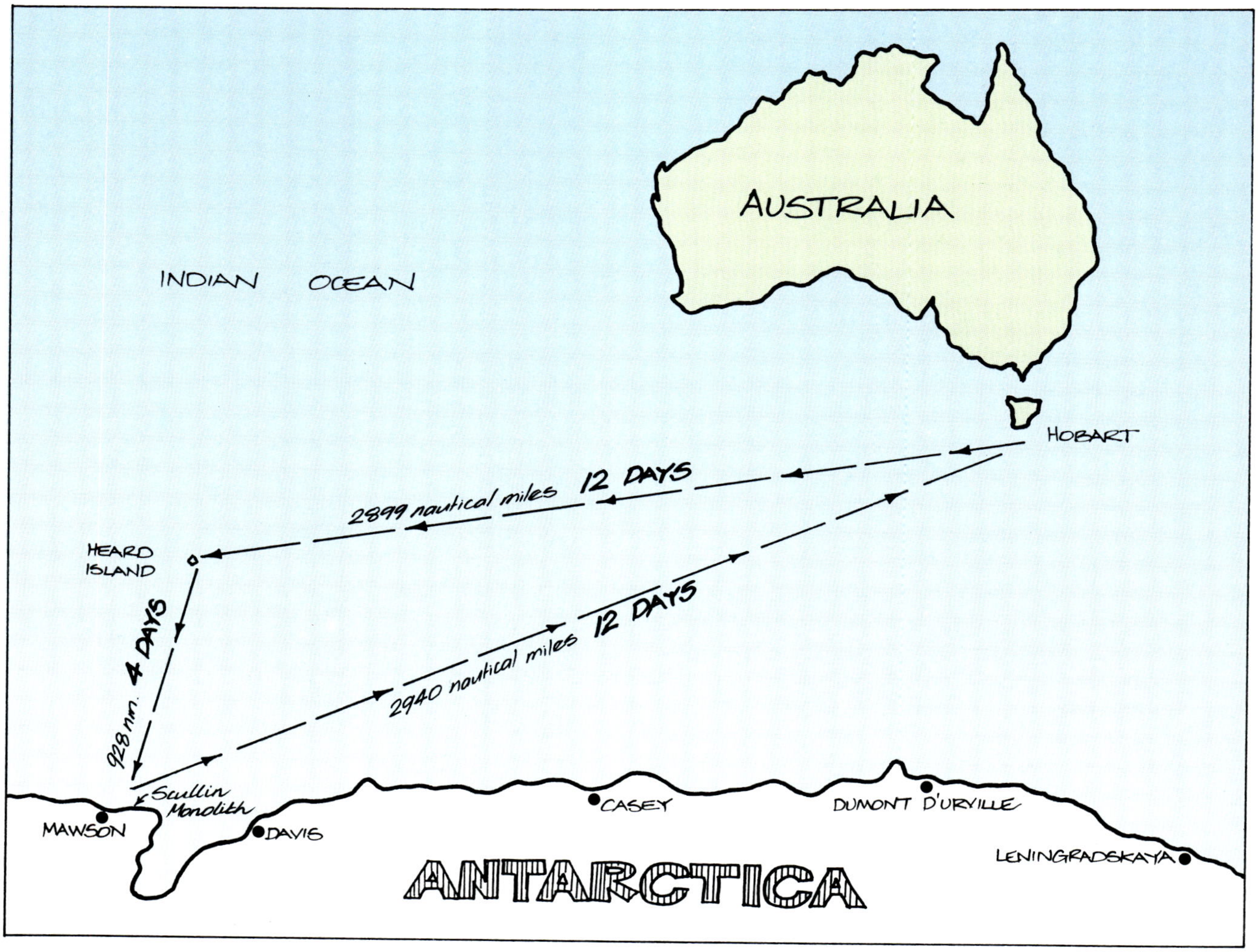
AUSTRALIA
INDIAN OCEAN
HOBART
2899 nautical miles 12 DAYS
HEARD ISLAND
928 nm. 4 DAYS
2940 nautical miles 12 DAYS
Scullin Monolith
MAWSON
DAVIS
CASEY
DUMONT D'URVILLE
LENINGRADSKAYA
ANTARCTICA

The broad expanse of the Amery Ice Shelf

Early in 1989, the capsize of the Argentine supply and cruise ship, *Bahia Paraiso*, caused the first major environmental crisis in Antarctica. The ship ran aground off Anvers Island on the Antarctic Peninsula, about 3 kilometres from the United States Palmer research station. Nineteen thousand litres of fuel from the ship's ruptured tanks began spilling into the sea. Its first impact was to kill krill, the small shrimp-like crustacean that is the linchpin of the Antarctic food chain. Krill were reported to be jumping out of the water. Penguins and other sea-birds were also affected, both physically, from direct contact with the oil, and behaviourally, presumably a response to eating the polluted krill. Twelve thousand sea-birds nested in the rookeries adjacent to Palmer station in 1988.

The ice-strengthened West German resupply vessel *Icebird* on charter to the Australian Antarctic Division resupplies Davis base in the Vestfold Hills.

Summer melt of Antarctic ice allows the growth of icicles.

Also on board the *Bahia Paraiso* was more than 600 000 litres of diesel fuel in barrels. Although not ruptured, the barrels floated away from the ship. Some were recovered, some were lost, some presumably sank. On the seafloor they have become petrochemical time-bombs creating a long-term threat to Antarctic wildlife. In cold Antarctic waters, oil takes one hundred times longer to break down than in temperate waters and the effects of the spill may be evident for one hundred years or more. This is the first serious pollution event in the Antarctic, but is it 'the tip of the iceberg'? How long can the Antarctic environment absorb catastrophes like this? Now that the Antarctic Treaty nations have paved the way for mineral and oil exploration during, perhaps, the next decade, to what extent will Antarctica be threatened by similar events in the future?

At all of the research bases in Antarctica there is the unmistakable mark of humankind trying to conquer the last remaining wilderness. There is no doubt that the accumulation of valuable scientific knowledge is the justification for the intrusion, but to what ends? It is too easy for these scientific bases to be regarded simply as foreign affairs outposts, demonstrations of sovereignty and control over what is still disputed territory, toeholds on what is becoming an increasingly valuable economic and perhaps strategic resource.

There is only one Antarctica. It is the last unexploited continent on earth. It is remote, unique. It can never be replaced, and more particularly, it could never be repaired. We, as a species, have a single opportu-

Strong katabatic winds armed with fragments of ice are capable of carving away the Antarctic ice surface to reveal centuries of accumulated ice and snow.

nity to preserve a whole continent in its natural state, to demonstrate we can leave the Last Great Wilderness on earth as wilderness, to show ourselves passionate, sensitive and caring about our planet.

Lichen Valley, Vestfold Hills

Lichens are one of the few simple plant forms that are found on the Antarctic continent.

Icebergs carved off the massive Amery Ice Shelf surrounded by summer pack-ice

PART I

A CONTINENT OF EXTREMES

Antarctica has the most extreme climate on earth. It is the coldest and windiest place on the globe. Minus 89.6°C has been recorded at Vostok, the Soviet base high up on the east Antarctic icecap. Along the coast the mean temperature in winter is between -15°C and -30°C; in the interior -40°C to -70°C can be expected. Mean temperatures inland in summer rise to the range of -35°C to -15°C. Temperatures seldom go above 0°C, although 7°C or 8°C is sometimes reached.

When the Australian scientist and explorer Douglas Mawson established his hut at Cape Denison, George V Land, in 1912, he called it 'the Home of the Blizzard'. The average annual wind speed was 80 kilometres per hour, with gusts in excess of 320 km/h. The wind that dominated conditions around Mawson's Hut was a katabatic wind, not the sort of wind we can deduce from the isobars on a weather map, but a locally produced wind. Because cold air is dense it can move down slopes in response to gravity. In Antarctica, very cold air on the inland plateau flows down, off the high land towards the coast in a shallow layer, perhaps 300 metres thick. By the time it reaches the coast it produces an almost constant gale.

Despite the low temperatures in Antarctica, there is very little annual precipitation. The coastal margins have the highest totals: the northern end of the Antarctic Peninsula receives the equivalent of about 900 mm annually, but on average, the mean annual accumulation is equivalent to less than 50 mm — a little more than in the Sahara Desert. That's the equivalent of just one good tropical thunderstorm, or three or four days' rain in temperate areas. There is no actual rainfall though, only snow, and very little of it. The enormous icecap has been built up by snowfalls deposited over millions of years.

Antarctica is the highest of all the continents, averaging 2300 metres

When the wind drops on clear Antarctic days, wind chill effect is reduced sufficiently to make for relatively mild conditions. While the sun creates a halo in the cirrus cloud overhead, a Weddell seal dozes under the summer sun.

above sea-level (North America by comparison is an average 720 metres). The highest peak in Antarctica rises to 5140 metres (16 859 ft) in the Vinson Massif in west Antarctica.

The continent of Antarctica is very old and some of its rocks, at 4.6 billion years, are the oldest on earth. It was once the central block of a large southern 'supercontinent' called Gondwanaland. Slowly, Gondwanaland split up into the continents as we now know them. Until just 40 million years ago — and that's fairly recently in geological terms — Australia and Antarctica were still neighbours. At about that time, Australia, like a giant iceberg itself, began a slow drift northwards to its present position.

Ice covers 98 per cent of the continent. It is the single largest body of freshwater or ice in the world. The ice is perpetually moving too. In the interior of the continent, where slopes are gentle, rates of movement are in the order of a metre or two per year. Closer to the coast, where slopes are steeper, the velocity of the ice movement may be 100 metres per year, and in an advancing glacier, ice can travel a kilometre annually.

Antarctica plays a key role in the energy balance of the globe and its climate affects us all in some way. The reason that Antarctica stays so cold is that its snow and ice cover reflects 80 per cent of the sun's shortwave radiation. Snow is also efficient at reflecting long wavelengths, causing the surface to cool further. For most of the year the centre of the continent actually loses more solar radiation than it gains.

This happens especially during the southern winter when Antarctica experiences twenty-four hours of darkness each day. At the height of summer the sun does not set. Despite the long hours of daylight, the continent does not heat up markedly because the incoming sun rays strike it at such a low angle. The great height above sea-level of the Antarctic Plateau keeps temperatures low too. The high interior is much colder than the coastal margins, but when temperatures drop in winter and the sea around Antarctica freezes, the coastal fringe effectively moves inland.

While being located at high latitudes keeps Antarctica cold, places at equivalent high latitudes in the northern hemisphere do not become as cold. The Arctic is not as cold as the Antarctic because it is not as big, and since there is no continental landmass in the Arctic, the ice there is close to sea-level.

Antarctica plays an important role in global weather patterns. The atmosphere and oceans of the earth are constantly circulating around the planet balancing the earth's temperatures. In contrast to the Antarctic, tropical areas absorb an excess of heat energy. Warm air from tropical areas near the equator slowly circulates southward and is cooled, and the cooler air is returned northward into the temperate zones. We see this cold, polar air as cold fronts on our weather maps. In this way Antarctica works like a giant air-conditioner.

The weather is the most important factor in Antarctic life. On the bases

Colours of the Antarctic sunset reflected in the milk-white walls of a tabular iceberg

The margin of the Forbes Glacier from the David Range in the Framnes Mountains

people call it the 'A-factor', the Antarctic factor. With extreme cold, drifting snow, and brutal winds, simple tasks can seem twice as hard and take three times as long to complete. It's a severe and unforgiving climate. The weather can mean the difference between life and death in Antarctica — something that we are still learning to cope with.

Man's inability to contend with the extremes of the Antarctic environment is reflected in the distribution of the scientific bases around the continent, and the extent and period of habitation. Ninety-eight per cent of Antarctica is covered by ice, but the scientific stations are almost all located on the limited number of ice-free areas along the very margins of the landmass. The United States of America's Amundsen–Scott base is

The Antarctic icecap rises behind the ice-free Vestfold Hills, Australian Antarctic Territory.

located strategically at the South Pole and the Soviet station Vostok is high on the east Antarctic icecap at 78°S latitude, but the rest of the stations represent only modest incursions onto the real Antarctic landmass. Expeditioners who have the opportunity to spend time in Antarctica are well aware that the real Antarctica is 'out there' and point south to the ice. Perhaps that is why the ice-free areas are referred to as 'oases'.

People are not natural inhabitants of Antarctica and the hostility of the environment requires the establishment of significant and sophisticated accommodation for any visit of an extended period. Still, none of the human inhabitants are permanent and few expeditioners stay for more than two years.

Exposed pre-Cambrian rocks of the Framnes Mountains south of Australia's Mawson station

THE ICECAP

Ice is the single dominating feature of Antarctica. The Antarctic continent is almost twice the size of Australia. It is a large landmass made up of extremely old rocks, pushed below sea-level under the weight of the ice that rests on top of it. The icesheet is thousands of metres thick. In one place in eastern Antarctica, in Wilkes Land, where the ice fills a deep trench carved in the underlying rocks by an ancient glacier, it is 4776 metres thick. Elsewhere the ice cover is thin, particularly where rocky mountain ranges push close to the surface. Isolated peaks of rock not covered by ice but surrounded by the relatively flat surface of the icecap are known as nunataks.

There is enough ice in Antarctica to bury Australia to a depth of 4 kilometres. As the rates of precipitation in Antarctica are very low, it is clear that the ice has accumulated over many thousands of years. The annual accumulation of snow and ice is revealed in the walls of crevasses in glaciers, icecliffs, and in the walls of icebergs. To study the layers of ice, scientists take deep cores from the icecap. By counting the layers of ice, each of which marks a season's snowfall, analysing oxygen isotopes, using carbon dating and a range of other techniques, they are able to gain information about the age of the ice and the nature of past climates.

Air bubbles trapped deep in the icecap are also recovered in the cores. The air in each bubble provides a tiny sample of the composition of earth's past atmospheres. Analysis of the air has provided valuable data about the amount of carbon dioxide (CO_2) in the atmosphere in past climates. This has become particularly important in our understanding of the Greenhouse Effect. As well as providing data on carbon dioxide, the product of the burning of fossil fuels, the bubbles in the ice cores have also allowed scientists to determine ancient concentrations of other important atmospheric gases such as methane and the relatively new chlorofluorocarbons (CFCs), which are also causing the destruction of the earth's ozone layer.

Effects in Antarctica are crucial in regard to both the warming of the atmosphere through the Greenhouse Effect, and the destruction of the ozone layer. Global warming due to increasing concentrations of atmospheric CO_2 has been recognized since the Industrial Revolution. In the last twenty years, the concentration of CO_2 has increased 25 per cent, due largely to the destruction of our tropical rainforests and the burning of fossil fuels. Global warming is now occurring perhaps one hundred times faster than at any other time in human history because the carbon dioxide and the methane and CFC levels are increasing so rapidly. Methane traps twenty times as much heat as CO_2, and a molecule of CFC is 20 000 times more efficient in trapping heat than a molecule of CO_2.

Studying ice cores has enabled scientists to verify the recent rapid increase in these 'Greenhouse gases' in the atmosphere. There are various estimates of how global temperatures will increase due to the Greenhouse Effect. Models predict an increase of between 1.5 and 4.5°C over the next fifty years. Equatorial regions are expected to warm the least; the increase in higher latitudes may be three to five times greater than in the equatorial regions. When you consider that when the earth's temperature was 5° colder than it is now, we were in the middle of an ice age, the problem becomes more obvious: a small rise in global temperatures would cause smaller bodies of ice like glaciers and pack-ice to melt, but the larger increase could melt the Antarctic icesheets.

Although the Greenhouse Effect may cause an increase in snow and ice accumulation because of higher precipitation rates, global warming is expected to cause the melting of the ice shelves at their bases, more rapid mobilization of the ice and greater rates of discharge to the oceans.

Antarctica is the largest body of freshwater in the world — 68 per cent of the world's freshwater is bound up in the ice. The first threat from Greenhouse warming is the melting of the west Antarctic icesheet. It is estimated that this would cause a global rise in sea-level of about 6 metres, enough to flood the majority of big cities in the world. If the east Antarctic icecap were to melt, world sea-levels would increase by an estimated 60 metres.

The parallel global problem of ozone depletion has equally important ramifications for Antarctica. Composed of three atoms of oxygen (O_3), ozone is not dissimilar to the oxygen (O_2) that makes up most of our atmosphere. Ozone itself is a poisonous gas; nowadays it is commonly found at the earth's surface over our big cities in smog, where it builds up as a by-product of the catalysis by sunlight of air pollutants from the exhausts of our motor vehicles. However, in the earth's outer atmosphere it plays a different role and is concentrated in a thin layer in the stratosphere, 16 kilometres above the earth. This ozone layer is only 50 kilometres thick but it is essential to the well-being of life on earth. It shields all living things from the sun's damaging ultraviolet rays. In fact it was the presence of the ozone layer in earth's atmosphere that enabled simple life forms to emerge from their protective aquatic environment, to evolve into amphibious creatures and subsequently into the life forms we are familiar with today.

Strong winds blowing across the top of the ice sculpt the surface to produce a pattern known as sestrugi.

Molecules of chlorofluorocarbon (CFC) in the atmosphere are not only very efficient at trapping heat, but they decompose, releasing chlorine atoms which break down ozone. Halons, used widely in the fire protection industry, are ten times more efficient than CFC in breaking down stratospheric ozone. This process is encouraged in the cold atmosphere over Antarctica during the three months of winter darkness, and the continuing destruction of the ozone layer has led to the development of an ozone 'hole' over Antarctica each spring. The 'hole' was first noticed by British scientists in 1981. It has since been verified by observations from pole-orbiting satellites, and is increasing in size each year. Antarctic ozone had been depleted by 50 per cent in 1987 and there have been

From atop a ridge of ice that has collected in the lee of Rumdoodle Peak in the Framnes Mountains, the Antarctic icecap reflects the slowly sinking sun. The ice ridge, created from accumulated blown snow and ice, is similar in shape to a desert sand dune.

losses of up to 95 per cent at some levels of the atmosphere. In the northern spring of 1988, after some speculation in previous years, an ozone hole over the Arctic was also found.

The destruction of the earth's protective ozone layer will expose us to biologically damaging ultraviolet radiation from the sun. The incidence of human skin cancer is expected to rise dramatically as a result, and plant and animal growth will also be reduced.

The effect of increased exposure to ultraviolet radiation in Antarctica is unknown. It seems that maximum levels of radiation do not coincide with the time of maximum phytoplankton productivity: the Antarctic ozone hole is at its maximum in early spring, well before the summer phytoplankton bloom which is so important to food supply in Antarctic waters. The effect on Antarctic animals in their breeding season may prove to be of even greater concern. The effect of increased exposure to ultraviolet radiation on the animals which come ashore to breed in spring and on their young is largely unknown.

Ice sculptures

The blue ice of the icecap is hard and very slippery, particularly after a day under the sun.

The Framnes Mountains

South of Australia's Mawson station are the Framnes Mountains. They are made up of the Casey Range to the west, the David Range and the North, Central and South Masson Ranges in the east, together with scattered nunataks, the smaller masses of rock that protrude through the icecap. The mountains are exposed, uncovered by the Antarctic icecap. These largely ice-free masses of rock are the only positive topographic elements on the icecap for hundreds of kilometres. As such they interrupt the dominant katabatic winds that flow down off the icecap towards the coast. Funnelled around the peaks, and armed with fragments of snow and ice, these winds are capable of carving the solid ice, leaving deep ravines. We found such a windscour near Fang Peak, in the David Range.

Walking into the windscour was like going down an icy staircase to a point well below the normal surface of the ice. On the walls of the ravine, where the ice had been carved away, the accumulation of ice and snow over the centuries was revealed in the layers. Once down inside the windscour, with the Antarctic panorama beyond view, the whole scene becomes more local and more intimate. The vastness of the Antarctic landscape is temporarily forgotten and there is an opportunity to be aware of the microfeatures of the ice itself.

The surface of the icecap is extremely difficult to walk on. If available, metal crampons should be strapped to the soles of your boots to reduce slipping, or you can use an ice-axe as a walking stick. It's like walking on the melting surface of a huge icecube — you can even see beneath the surface of the ice into the blue ice below — or like walking on water over a bottomless blue abyss.

Walking on the broken rocks at the foot of these ranges of the Framnes Mountains provides another aspect of this fascinating place to appreciate. It is different in so many ways from most other continents on earth. The rocks produce virtually no soil and there are no plants to speak of. Despite that, there are sea-birds circling overhead, 30 kilometres from the coastline. In fact pairs of beautiful pure-white snow petrels are nesting in crevices in the rock in these mountains, high above the icecap.

Strong winds laden with ice fragments have carved away the icecap adjacent to the rocky ridges near Fang Peak.

Television crew in the windscour

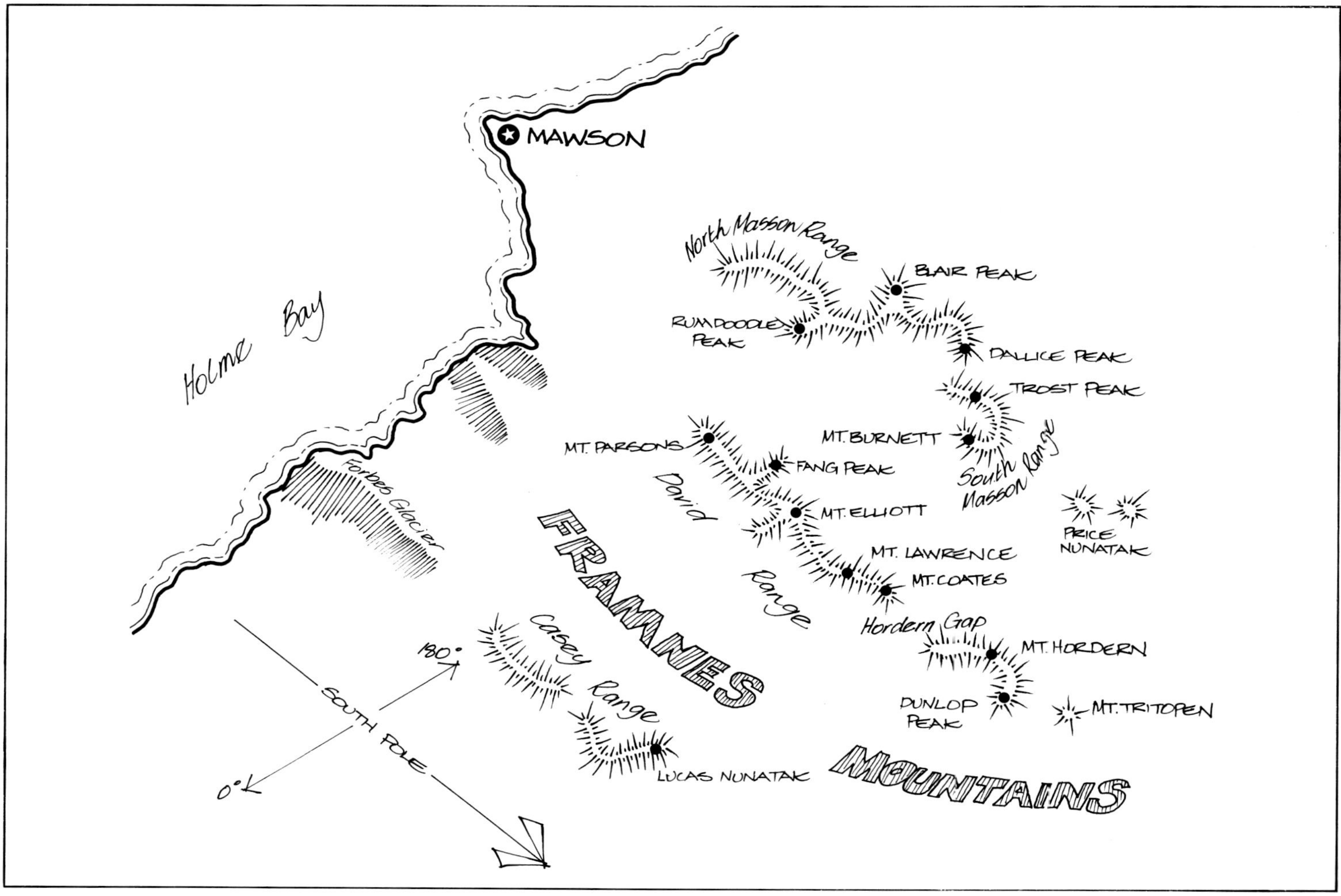

To have the opportunity to spend a night inland, away from the coast, if only a relatively short distance onto the icecap itself, is a remarkable experience. With the summer sun just brushing the horizon at its lowest point in the sky just before midnight, the glistening ice after a sunny day reflecting the colours of that midnight sun, and no wind, there is a sense of majesty not found in many other places. With no trees to rustle in the breeze, there is silence, a silence so complete that it is necessary to move, to create noise as reassurance that you can still hear.

This single boulder carried in a glacier has been exposed after the surrounding ice has melted away. The boulder now retains the radiant energy from the sun and is causing accelerated melting at its base.

Antarctic sun over the icewall in the windscour

Scullin Monolith

On the coast of Antarctica about 250 kilometres east of Mawson lies a large exposed rock monolith named after the Australian prime minister James Scullin. The rock is Precambrian gneiss and among the oldest rocks on earth.

Scullin Monolith is the site of the largest colony of Antarctic petrels in the Prydz Bay region. In the summer of 1987 Australian scientists estimated the population at 158 000 breeding pairs. These beautiful birds make a small nest of stones high up on the cliff face and raise one chick. Nests may be within a metre of each other. Antarctic petrels are one of the handful of bird species that breed on the Antarctic continent itself. Most species in this part of the world choose sites on the sub-Antarctic islands where food supplies are more plentiful.

Seldom are both parents seen together at the nest; mostly they are at sea, feeding. They are masters of flight, and like other Antarctic birds they rest infrequently, although they are sometimes seen in large flocks on icebergs.

Cold, dense air drains downslope from high on the Antarctic Plateau to produce strong katabatic winds at the coast. At Scullin Monolith the wind blows loose snow across the ridges. In their lee, Antarctic petrels nest beneath the brow of the cliff and Adélie penguins colonize the rocky slopes beyond the wave zone.

From the sea Scullin Monolith seems small but Mikkelsen Peak on the left rises 420 metres above the sea. The Antarctic icecap is plastered against the landward side of the monolith.

Shattered boulder on Scullin Monolith

From a distance this iceberg appears stained. Closer inspection reveals that the brown coloration is in fact thousands of Antarctic petrels at rest from feeding.

Antarctic petrel and chick on Scullin Monolith

AMERY ICE SHELF

High up on the Antarctic Plateau, snow and ice accumulate very slowly on the surface. What we now see as the massive amount of ice bound up in the east Antarctic icesheet has taken millions of years to build up. At one point on its margin, that ice moves slowly down the Mawson Escarpment under gravity, becoming the Lambert Glacier, the biggest glacier in Antarctica. Where the Lambert Glacier reaches a giant fjord at the coast, the ice moves out over the sea forming the vast Amery Ice Shelf, almost 200 kilometres across. The Amery Ice Shelf is a huge layer of slowly but constantly moving, floating ice, continually fed on its landward side with ice from the Lambert Glacier.

From high in the air, the Amery Ice Shelf looks like a thin layer of ice. But from a low-flying helicopter, the enormous size of the icesheet becomes apparent, another reminder of the difficulty of comprehending the scale of landforms in the cold, dry Antarctic air. Very often distances are compressed, mountain ranges seem much closer than they really are.

As one of the largest ice shelves in Antarctica, the Amery is a major source of icebergs in eastern Antarctica. They break off the seaward edge of the ice shelf and slowly move away into Prydz Bay and the Southern Ocean. This process is called 'calving'. Many of the large, newly formed icebergs from the Amery Ice Shelf become grounded on Frome Bank near Cape Darnley to the west.

The Filschner Ice Shelf is another major shelf, but the largest of all is the Ross Ice Shelf which is about the same size as France.

In summer the pack-ice forms only a thin fringe adjacent to the ice-cliffs, and only the larger pieces of pack-ice remain. However, even at the end of the summer, in February, about 3 million square kilometres of the sea are still covered with ice. The sea-ice starts developing again towards the end of March, and continues to grow through the dark hours of winter. It reaches its maximum extent in September or October. At this time the sea-ice may form a barrier 500 kilometres wide and up to 2 metres thick, and extend over an area of about 20 million square kilometres. The area of sea-ice that builds up during the winter is larger than the Antarctic continent itself.

A broad expanse of pack-ice with the enormous floating Amery Ice Shelf beyond

The pack-ice serves an important function in keeping Antarctica cold. As the sun rises higher in the sky through the spring — the same time that the sea-ice is at its maximum extent — the ice reflects about 80 per cent of incoming solar radiation and keeps the air cold. In comparison the open ocean reflects only about 5 per cent of solar radiation.

Sea-ice actually lowers the temperature of the air above it, and when it melts it reduces the temperature of the seawater. Both of these facts have important consequences for atmospheric circulation, for the temperature of the Southern Ocean, and hence for the climate of the Southern Hemisphere.

Through the summer months the pack-ice is broken and reduced in extent. From the air it resembles a terrazzo pavement surrounding the island icebergs.

The seaward margin of the Amery Ice Shelf

It's not until the detail and height of the icecliff are revealed that the full size of the Amery Ice Shelf is appreciated.

ICEBERGS

Icebergs are perhaps the most fascinating of all the physical features that are unique to Antarctica. They are all individual. Some are angular, some are smoothed by water, having split and toppled over. Most icebergs live for about four years before melting, though larger ones can survive for longer periods. An average iceberg measures a few hundred metres in width, and is perhaps 750 metres tall. We see only as far as the blue shadows below the water-line — 90 per cent of these leviathans lies beneath the surface. In shipping parlance, they fall into various categories according to size: there are the big 'bergs' the size of a city block; 'bergy bits' about the size of a house; and 'growlers', pieces about the size of a grand piano that float just beneath the surface — they're called 'growlers' because that's the noise they make as they slide along a ship's hull.

Icebergs are also described according to their shape. The flat-topped ones are said to be tabular, like a table-top. The more angular ones are described as irregular, and those that have been washed and smoothed by the sea, then toppled over to expose a smooth surface, are said to be rounded bergs.

Large icebergs may be 5 kilometres long or more, and the height of a ten-storey building. In 1967 a giant tabular iceberg broke off the Amery Ice Shelf. It was more than 100 kilometres long and 50 kilometres wide; it was tracked on weather satellites for three years and drifted more than 4000 kilometres before breaking up. More recently a giant iceberg broke off the Filschner Ice Shelf — it was almost half the size of Tasmania.

We spent one night in *Icebird* cruising between hundreds of these icebergs on Frome Bank between Mawson and Davis stations. Because most of an iceberg lies beneath the water, in the relatively shallow water of about 200 metres on the continental shelf, larger icebergs can easily become grounded on the seafloor. This provides an ideal opportunity to move closer to the walls of the icebergs. In many cases you can clearly see the layers of ice that have built up over millions of years; these layers are like the rings of a fallen giant tree in a forest. The exposed profile of a

The pastel colours of an Antarctic sunset are enhanced when reflected in the glistening wall of an iceberg.

Tabular icebergs vary in size from a few hundred metres to many kilometres across. Most of their bulk lies below the surface, making them, for all intents and purposes, immovable.

large iceberg might represent the period of the evolution of man.

A number of nations, including Australia, are investigating icebergs as a source of freshwater. It is thought that a medium-sized iceberg could be selected in Antarctic waters, wrapped in plastic and then towed through temperate waters to provide enough water for a small city for one year — perhaps $25 million worth of water at today's rates.

Cave in the seaward edge of the Sorsdell Glacier

Relatively small icebergs like this one are referred to as 'bergy bits', but most are the size of a large house.

Fractures in large icebergs reveal the deep blue of the freshwater ice bound up inside them.

As large icebergs move away from Antarctica with oceanic currents they become exposed to the strong westerly winds of the 'Furious Fifties'. The largest survive for many years before breaking up.

Annual layers of ice and snow built up on the Antarctic plateau are moved to the coast in glaciers and icesheets, then broken off in segments to form icebergs. The compressed time profile represented by these layers could mark the period of the ascent of man.

HEARD ISLAND

Unlike our bases in Antarctica, tiny sub-Antarctic Heard Island and its near neighbours the McDonald Islands are Australian sovereign territory. In fact this is our only sovereign territory in this part of the world. These islands became part of Australia in 1950 after an 'exchange of letters' with the government of the United Kingdom. Heard Island is a tiny speck of land in the Southern Ocean, 4100 kilometres southwest of Perth and only 42 kilometres long. Although perhaps smaller than the city of Canberra, at its highest point it is 600 metres higher than Mt Kosciusko. Heard Island is Australia's only active volcano and last erupted in January 1985. At more than 2800 metres (9200 ft), Mawson Peak on Big Ben is by far Australia's highest mountain.

The slopes of Big Ben are glaciated to the sea, but rocky promontories at the foot of the mountain provide breeding sites for a fabulous array of Antarctic birds and animals. At latitude 53°S, the island lies very close to the Antarctic Convergence Zone, the biological boundary between the cold Antarctic waters and the warmer sub-Antarctic waters to the north. Upwelling of nutrients from the deep ocean at this point provides abundant food supply and correspondingly large populations of penguins, other sea-birds and seals breed on the sub-Antarctic islands like Heard.

Heard Island was first sighted by the British sealer Peter Kemp in 1833. It was named after the American Captain John Heard in 1853. The remnants of a stone-walled American sealers' hut built in 1860 still remain. Fur seals were almost exterminated for their skins, and populations of elephant seals have only slowly recovered after the slaughter for the oil in their blubber, boiled down in tripots over open fires.

Heard was visited in 1874 by HMS *Challenger* on her scientific expedition for the Royal Society, and in 1902 by the *Gauss*, a German ship under the direction of the geographer Erich von Drygalski. Douglas Mawson visited Heard in 1929 as part of the BANZARRE expeditions on *Discovery*. Despite the furious winds of the fifties latitudes, Heard Island became the site of the first Australian National Antarctic Research Expedition in 1947 and the base there was maintained until Australia's first Antarctic mainland base was established at Mawson in 1954.

The snow and glacier covered peak of Big Ben, Australia's only active volcano, emerges from the almost permanent shroud over Heard Island.

Evening sun on Big Ben

Glaciated slopes on Heard Island. Across the world glaciers are melting and retreating.

The exposed volcanic lavas of Heard Island provide an ideal habitat for the thousands of sea-birds that feed within the zone of the Antarctic Convergence.

Young male elephant seals bask in the sun.

Heard and McDonald Islands (a small group of rocky islets nearby on which there are large breeding colonies of sea-birds) have fortunately been placed on the Register of the National Estate. Despite its bloody history during the sealing era, Heard remains relatively unspoilt. These are the only sub-Antarctic islands where no introduced animals or plants grow, with the exception of one species of grass, *Poa annua*. In comparison, Macquarie Island, Australia's other sub-Antarctic island and part of the state of Tasmania, has suffered badly from the introduction of rabbits, cats, mice and black rats. There is a strong case for Heard and McDonald Islands to become World Heritage areas.

The seal populations are still recovering from the intensive sealing of the last century, but southern elephant seals and Kerguelen fur seals are still found in abundance during the breeding season, and leopard seals are also found.

Birds are abundant on Heard Island. King and gentoo penguins are resident on Heard and McDonald Islands and their breeding colonies are enormous. Chinstrap, macaroni and rockhopper penguins also breed on these islands, as do colonies of terns, storm petrels, diving petrels, prions, fulmars and albatrosses.

Another unusual bird is the sheathbill. These are true land birds; pigeon-sized, they are white with leathery facial wattles. There are two species, one restricted to the Antarctic Peninsula; the other, the lesser sheathbill, is found only on the Indian Ocean sub-Antarctic islands like

Elephant seals, probably young males without established harems of females, play in the shallow bays of Heard Island. As adults they will use the techniques they learn in these games to vigorously defend their territory and harems from rival bulls.

Heard and Kerguelen and remains on the islands through the winter. On Heard Island it is seen scavenging around penguin colonies or seal wallows, and often around the buildings of Atlas Cove base.

A relic of Heard's bloody sealing era; a sealer's knife still lies on the volcanic sand.

The remains of a stone-walled sealer's hut built in 1860 on Heard Island

THE SOUTHERN OCEAN

Separating Antarctica from the other continents of the world is the vast Southern Ocean. Most of the Southern Ocean lies south of 40°S latitude, embracing the southern sections of the Antarctic, Indian and Pacific Oceans. Through history it has been an enormous physical barrier to the explorers, and it still can be a barrier to the Antarctic continent for those approaching by ship. Mountainous seas can be whipped up very quickly by the wind, there is a strong circumpolar current and the danger of floating ice in the southernmost latitudes often makes shipping hazardous. Very few ships use the waters of the Southern Ocean apart from those servicing Antarctic bases, the international fishing fleets, and a handful of tourist vessels and private yachts.

Conditions in the Southern Ocean, particularly weather conditions, make it a cold and inhospitable place. The strong and unbroken westerly windstream across the latitudes of the Southern Ocean has led sailors to describe its waters progressively southward as the 'Roaring Forties', the 'Furious Fifties' and the 'Shrieking Sixties'. It is the Southern Ocean that transmits Antarctic weather to the southern continents like Australia and South America.

Water circulation in the Southern Ocean responds to the dominant winds. In the northern part, the prevailing West Wind Drift sets up an eastward-moving current. The water nearest the continent circulates in the opposite direction in response to the dominant East Wind Drift. Apart from these circumpolar currents, the surface-water near Antarctica tends to drift north.

A very important physical feature in the Southern Ocean is the Antarctic Convergence Zone. This is a biological boundary formed where that cold Antarctic surface-water from the south sinks beneath warmer, southward-moving water from lower latitudes. The many small islands of the Southern Ocean that lie within the Antarctic Convergence reflect the temperature boundary. Those lying south of the convergence are glaciated, while those to the north are ice-free with no permanent snow cover.

Viewing the sea-birds of the Southern Ocean is one of the pleasures of the long sea voyage to Antarctica. Wandering albatrosses like this one will follow ships for many kilometres.

The convergence is less than 50 kilometres wide, and is found between latitudes 50° and 60°S, but it forms the northern boundary to the Antarctic region and is critical to the balance of nature in Antarctica and the Southern Ocean.

Southern Ocean

The Antarctic Food Chain

Animal life in the Antarctic environment is unique. In contrast to northern polar regions, there are no polar bears or walruses, but a vast array of whales, seals and birds, particularly penguins. A critical factor in the distribution of Antarctic animals and birds is the Antarctic Convergence, where cold Antarctic waters meet cool temperate waters, and where the Southern Ocean becomes rich in nutrients and abounds with plankton and krill.

An increase in nutrient salts occurs south of the convergence and the Antarctic surface-waters are much more productive than the sub-Antarctic surface-waters north of the convergence. Also, close to the coast of Antarctica is a zone of upwelling of nutrient-rich water which contributes to luxuriant growth of phytoplankton adjacent to the continent. This phytoplankton, minute plants that float and drift in the sea, is the first link in the Antarctic food chain.

There are about one hundred species of Antarctic marine phytoplankton, and they in turn provide food for the tiny floating animals known as zooplankton. Krill is the dominant organism of the Antarctic zooplankton. It is a small shrimp-like crustacean that is commonly found in huge swarms covering hundreds of square metres. It is the most abundant animal in the world. Krill forms the staple diet of many of the whales. As well, fish, squid, octopus, seals and birds all depend on krill and the other zooplankton for food. The krill, though, is the most important animal in the Antarctic food chain.

All of the nutrients, plants and animals of the Antarctic form an ecological network in a delicate balance, which is independent of other world ecosystems. Perhaps the key to the Antarctic food chain is its relative simplicity; but despite this simplicity, very little is known about it. There are two important parameters that make the Antarctic marine ecosystem unique.

First, species numbers in the Antarctic are low; there are not many different types of animals. Of the penguins, only emperors and Adélies actually breed on the shores of Antarctica; another five species are found

Krill refers to many species of planktonic crustacea. The dominant species is *Euphausia superba*, the Antarctic krill.

A krill swarm breaks the surface, possibly pursued by fish, penguins or seals.

on the peripheral islands, and on the sub-Antarctic islands near the Antarctic Convergence.

Of the other birds, only eight different species breed on the Antarctic continent or the continental islands. Most inhabit the islands to the north, away from the pack-ice and where winters are less severe. By way of comparison, different bird species in temperate or tropical regions of the globe number in the thousands.

Similarly with the other animals: there are only six species of baleen whales, and six species of toothed whales and dolphins commonly seen in the waters around Antarctica. There are four seals that inhabit the Antarctic coast and pack-ice; elephant and fur seals, the two that were ruthlessly hunted by man, are found mostly on the sub-Antarctic islands.

There are also a few species of fish, crabs, and squid, octopus and other molluscs.

Second, although species numbers in Antarctica are low, populations of each species of most Antarctic animals are enormous — there are millions of crabeater seals, for example. But because there are so few different species, the balance of nature is exceedingly fragile.

The impact of harvesting seals and whales to the point of extinction in the past has been disastrous. There is no doubt that exploitation of the baleen whales influenced the Antarctic marine ecosystem. It has been estimated that krill consumption by whales fell by about 75 per cent after the reduction in the whale populations. However, the populations of the other krill-consumers, seals and penguins, have probably increased in response. There is now evidence that commercial fishing fleets have severely depleted the numbers of some fin-fish. Krill is also the subject of commercial exploitation, and it is possible that already harvesting has had a significant impact on the krill swarms and that repercussions have already been felt by the animals that normally feed on the krill.

The other significant effect that mankind has had on the Antarctic ecosystem is through the dispersal of toxic chemicals from the industrialized continents to the north via the oceans and air. The discovery of DDT in Antarctic penguins shocked people into realizing the devastating ability of these pollutants to disperse around the globe.

It is vital that we find out all we can about Antarctic ecosystems. Armed with that knowledge, we can take measures to ensure that human activity does not endanger the survival of Antarctica's natural inhabitants or their environment.

Whales

In the Antarctic summer of 1930/31 alone, the whalers slaughtered an estimated 30 000 blue whales. Of this, the largest animal that has ever lived on earth, only a few thousand are now left and populations may never recover. As well as blue whales, the ships hunted sperm whales, humpback whales, right whales, fin and sei whales. A few of these populations are now recovering. Only minke whales have maintained anything like their original numbers — they are small and were uneconomical for the whalers to kill. Despite a recent international ban on whaling, minke whales are now being slaughtered. The Japanese take 857 annually for so-called scientific purposes, although very few scientists would argue that you can learn more from a dead whale than a live one. Whalemeat often ends up in the exclusive specialist restaurants of Tokyo.

Whales do not breed in Antarctic waters, they use the productive southern waters as a feeding ground. They feed to produce the thick layers of blubber that will sustain them through the breeding season in less rich but warmer waters. The minke whale, the smallest of the Antarctic whales, is the exception and spends the whole year in the Antarctic zone.

The whalers slaughtered the whales for their blubber, a subcutaneous layer of fat that can be up to 20 centimetres thick in the giant blue whales. The blubber provides the whales, and most of the seals, with an insulating layer that protects them from the cold of the Southern Ocean.

Minke whales in open pack-ice

Minke whales average about 8 metres and live for about 50 years. Their numbers in Antarctic waters may be increasing.

Seals

The seals are perhaps the most captivating of the animals encountered in the Antarctic. Well adapted to the aquatic environment, these animals still need to return to land to breed. The Antarctic supports a much larger number of seals than the Arctic because feeding areas are more productive and there are no native predators.

Crabeater seals are often seen lying on the icefloes. Despite their name, 90 per cent of their diet is krill together with some fish and squid. Crabeaters, with a population of perhaps 40 million, account for about half the world's seals. With such population numbers they consume more krill than the baleen whales.

Elephant seals are the largest of all seals and take their name partly from their size and partly because of the long trunk-like proboscis that the male can inflate when aggravated or alarmed. Adult males reach a length of 4.5 metres and a weight of 4 tonnes. Females are much smaller. They are not strictly an Antarctic species since they breed on the sub-Antarctic islands where they group in harems of up to 70 females. Dominant males with large harems are called 'beachmasters'. They will fast for three months over the summer breeding period while vigorously defending their territory and harem. Many are severely battle-scarred from wounds inflicted by the strong canine teeth of their rivals. Hunted mercilessly in the 1800s for blubber, populations have generally recovered well. Current estimates put their numbers at about half a million.

On the Antarctic mainland they haul up to moult and bask in the summer sun. Near Davis base in Australian Antarctic Territory, we watched scientists tagging a group of young males. Putting a cattle tag into the back flipper of these enormous seals is quite a task. Enjoying the warmth of the sun, and a day of about 2°C, they really didn't want to move and had to be coaxed apart to expose their flippers so that the scientists could either read an existing tag or put a new one in place. It is hoped that this tagging program will provide information about the population movements and dynamics of these seals.

Recent scientific data from this project indicate that southern elephant seals have undergone a drastic, unexplained population decline. Colonies at Macquarie and Heard Islands, two major sites, have seen numbers reduced by 50 per cent over a forty-year period. Populations at three of the remaining six major breeding centres around the world have also declined. Many researchers believe that over-fishing by commercial enterprises may be the cause of the decline.

Weddell seals inhabit the fast-ice along the coast of Antarctica. They

Young fur seal pup at Heard Island

Weddell seal near Mawson base

Leopard seal at Heard Island. These seals are strong and are fast swimmers. They have large heads and huge lower jaws, and can open their mouths very wide.

are often seen by visitors, and are often in large groups. Mature adults weigh up to 400 kilograms. They are the best swimmers of all the seals. They dive very deep and their eyes are specially adapted to low levels of light intensity. Deep dives of up to 600 metres have been recorded and the seals can remain underwater for more than an hour. Through the dark of winter they remain beneath the ice, maintaining breathing holes by scraping the ice from below with their teeth. Their diet consists mainly of fish, but squid and other crustaceans are also taken.

The Ross seal is less often seen by visitors; it is the least abundant of the Antarctic seals, and is the least well known. Ross seals inhabit the consolidated pack-ice and feed mainly on squid.

The largest of the true Antarctic seals is the leopard seal. Leopard seals are most often seen alone but are widespread through the Antarctic. They have a large head and strong jaws and are most often seen cruising the fringe of a penguin rookery. Their diet consists of almost equal quantities of birds, mostly penguins, young crabeater seals and krill.

The Antarctic fur seal is not a true seal. It has a visible earflap and is therefore related to the group that includes the sea lions. Fur seals do not have the same insulating layer of blubber as the true seals. Instead their thick coats trap a layer of air which provides thermal insulation. They also use the exposed skin of their flippers to help regulate their body temperature, losing heat by radiation away from their bodies, or conserving heat by tucking the flippers under. These flippers are very strong and fur seals use them to move about quite well on land. Claimed for their skins in the nineteenth century, they almost disappeared. With protection, their populations have now largely recovered.

Birds

Bird life in the Antarctic is quite bewildering; Antarctica provides habitat for an estimated 100 million breeding birds each year. Of those, most cluster on the tiny sub-Antarctic islands close to the Antarctic Convergence. Forty-three species of birds breed within the limits of the Antarctic Convergence where food is readily available; very few species breed on the Antarctic continent itself. By far the most numerous are the petrels, ranging from the tiny Wilson's storm petrel weighing only 35 grams to the wandering albatross with a wingspan of 3 metres.

The petrels all have 'tube noses', possibly to assist in draining away concentrated brine from a gland above the eyes. They are all marine species, feeding at sea, the smaller species taking zooplankton and krill, and the larger ones, in particular the albatrosses, large squid and fish.

Giant petrels, like all petrels, lay only one egg each season. They come ashore to nest and breed on a handful of rocky sites. Chicks at one breeding site near the Australian Davis base are banded annually to monitor population numbers and migration patterns. Without their flight feathers their only defence is to aggressively vomit a mixture of specially concocted red oil along with the remainder of their last 'fishy' meal — quite a deterrent.

Antarctic petrels also uncharacteristically breed on the Antarctic continent rather than the sub-Antarctic islands further north. They compete with other birds and mammals, including man, for the few ice-free sites available. They feed at sea on plankton and krill, and are often seen in large numbers over the pack-ice or resting on icebergs.

The penguins are by far the most numerous of Antarctic birds and they are the birds that most people associate with Antarctica. The most majestic are the emperor penguins which stand a metre tall. They nest in colonies of up to 80 000, but interestingly it is the males that incubate the single egg. They huddle together through the dark and cold of winter, each with an egg resting on his feet, a fold of belly skin over the egg further protecting it from the cold.

Fifty per cent of Antarctic penguins are Adélies. They, like most other Antarctic birds, breed in the spring and summer, returning to the same nesting site and the same mate. The Adélies are first birds ashore in spring, scampering across the sea-ice to re-establish their nesting sites, and to begin the struggle to retain — or steal — the ring of stones that makes up the nest for their eggs. Adélies are seemingly ubiquitous through the summer months — total numbers are in the tens of millions.

All the birds that inhabit the Antarctic zone have adapted to the special conditions that are found there. Sea-birds, by their numbers, have obviously adapted better than land birds. They are able to conserve body heat because of their waterproof plumage, a layer of fat just under the

Giant petrel with chicks. Despite nesting on an exposed rocky site these birds are remarkably well camouflaged.

Adélie penguin rookery. Most of these birds are down-covered chicks waiting in crêches for their parents to return from sea after feeding.

Adélie penguins stand about 70 centimetres tall. They have extraordinary homing abilities and it is thought that they navigate by the sun.

Emperor penguins are the largest of all and unusually they breed in winter on fast-ice. This means that these birds almost never set foot on land.

Southern skuas are the scavengers of the Antarctic. They will take chicks from other bird colonies and are often seen around bases looking for food scraps.

skin, their compact shape and their lack of bare skin.

All Antarctic birds are highly vulnerable to the activities of man. Nesting birds, even penguins in dense rookeries, are easily disturbed and will desert their nest. To protect some of the larger rookeries, for example the Taylor emperor penguin rookery near Australia's Mawson base, access has been limited, usually to scientists only. However, many other sites, particularly those adjacent to coastal research stations or where tourists can easily come ashore, could be severely disrupted in the future unless steps are taken to guard these delicate communities.

PART II

HISTORY OF DISCOVERY

Antarctica was for centuries an unknown land. The notion of a massive southern continent was first postulated by the ancient Greeks, and indeed it was Aristotle who gave Antarctica its name. His theory was that the northern hemisphere continents must be balanced by a southern one. Since the region of the northern pole lay under the constellation of the Great Bear, *arktos* in Greek (hence 'Arctic'), Aristotle postulated that the southern equivalent of *arktikos* must be *antarktikos*, its opposite.

All of the early discoverers back to the sixteenth century, including Magellan and Francis Drake, were aware of the speculation about a 'southern continent'. The Falkland and Shetland Islands were discovered around 1600. In 1739, the French navigator Jean-Baptiste Charles Bouvet de Lozier landed on what he thought was the Antarctic mainland. It was in fact a tiny, isolated island, today reckoned to be the most remote piece of real estate on earth, southwest of the Cape of Good Hope. It is now known as Bouvetoya. In 1772, another Frenchman, Yves-Joseph Kerguélen-Trémarec, in search of the large southern continent, claimed to have found it. Returning in 1773, he was embarrassed to find only islands, the Îles Kerguelen.

After his discovery of New Zealand and the east coast of Australia in 1770, James Cook speculated about a great continent to the south. On his second major voyage in 1772 aboard the *Resolution*, he was instructed to circumnavigate the globe at high latitudes and voyage as close to the South Pole as he could. He was subsequently probably the first man to cross the Antarctic Circle, but he failed to sight Antarctica and concluded that there was no southern continent. Future explorers were discouraged but, as Cook had recorded large numbers of seals and whales in the Southern Ocean, European and American whaling fleets ventured south. They became the new explorers while at the same time beginning the first period of exploitation of Antarctica.

It was not until 1820 that Thaddeus von Bellingshausen, under orders from Tsar Alexander, first sighted the Antarctic Peninsula. Also in that year, the British naval captain Edward Bransfield made the same claim, as did the American sealer Nathaniel Brown Palmer.

With the door opened, the 1830s and 1840s saw the British sealers John Biscoe, James Weddell, Peter Kemp and John Balleny begin to chart the Antarctic coast. Then followed discoveries by the Frenchman Jules-Sébastien-César Dumont d'Urville, the American Charles Wilkes and British naval captain James Clark Ross, all intent on making observations of the South Magnetic Pole. At the same time, the sealing and whaling ships moved further south. In the early part of the nineteenth century, seals and whales were slaughtered in their thousands for the oil in their blubber. The oil was used as fuel, in particular for lamps. Whalebone was used for ladies' corsets.

By the middle of the nineteenth century whaling in southern oceans declined and only when the Norwegians developed gun harpoons with explosive heads did it again become economically attractive. It was because of the whalers, particularly the Norwegians, that scientific interest was rekindled in Antarctica.

After centuries of exploration and discovery, a confirmed landing on the Antarctic mainland was finally made in January 1895. A young Norwegian settler from Melbourne, Carstens Borchgrevink, a hand on board the Norwegian ship *Antarctic*, jumped ashore from a rowboat. There was later some dispute whether it was Borchgrevink or a New Zealander named von Tunzelman who was the first to step out of the boat, but their discovery of some lichen and some geological specimens, together with their geographical records, spurred on scientific exploration.

New scientific expeditions quickly followed. Borchgrevink led the British Antarctic Expedition 1898–1900, the first group to overwinter on the mainland. Robert Scott made his first attempt to reach the South Pole in November 1902. Ernest Shackleton almost reached the Pole in 1909. On that same expedition were the Australians Edgeworth David and Douglas Mawson. They were the first to climb the now infamous Mt Erebus; they also reached the South Magnetic Pole. On his second expedition Scott reached the South Pole on 17 January 1912, only to find that the Norwegian Roald Amundsen had been there one month earlier. Scott and his companions all died on the return journey. Mawson had organized his own expedition to Commonwealth Bay in 1911. Mawson's lone journey back to his base camp at Commonwealth Bay after the death of Belgrave Ninnis and Xavier Mertz made him a hero of polar exploration.

"RACING WITH DEATH IN ANTARCTIC BLIZZARDS"

ILLUSTRATED LECTURE

By SIR DOUGLAS MAWSON, K. B., D. Sc., B. E.

DR. X. MERTZ

One of Sir Douglas Mawson's comrades on the sledge journey, who lost his life from exposure and exhaustion.

SIR DOUGLAS MAWSON

Leader of the Expedition and the sole survivor of the tragic sledging journey.

LIEUT. B. E. S. NINNIS, ROYAL FUSILIERS

Sir Douglas Mawson's other comrade, lost by falling into a crevasse with his sledge and dogs.

In this Lecture Sir Douglas Mawson tells the story of his thrilling experiences in the Antarctic regions, in the course of which both his companions lost their lives, and he himself, after perhaps the most terrific perils out of which any adventurer ever escaped alive, was just able to reach safety. The lecture is fully illustrated with the most remarkable Still and Motion Pictures ever taken in Polar Regions.

Exclusive Management of LEE KEEDICK, 437 Fifth Avenue, New York

On his return from Antarctica in 1914, Mawson toured the world presenting his illustrated lectures, in part to pay for the additional cost of having to stay in Antarctica for an extra year after his ill-fated sledging journey with Ninnis and Mertz.

Douglas Mawson at Proclamation Hill, King George V Land, on 5 January 1931. Much of what has become Australian Antarctic territory was proclaimed British Sovereign Territory during the BANZARRE voyages of 1930–31.

Mawson's Hut, Commonwealth Bay, 1912, the so-called 'Home of the Blizzard'. This photograph by Frank Hurley, perhaps Australia's most noted polar photographer, was called 'Whetter and Close collecting ice for domestic purposes'.

The first Australian National Antarctic Research Expeditions (ANARE) base was set up at Atlas Cove, Heard Island, in 1947. The original buildings, including Admiralty Hut, are still standing and original supplies still line the shelves.

Old timbers from the early base at Atlas Cove are now overgrown with mosses.

THE ANTARCTIC TREATY

The expeditions of Scott, Amundsen, Shackleton and Mawson made the first decades of the twentieth century a period of fame, but notions of conquest and sovereignty were to follow. In 1923 Britain claimed the Ross Sea as a dependency in addition to the Falkland Islands. In 1924 the French claimed a slice of Antarctica immediately to the south of Australia. The Norwegians claimed territory in 1928. Douglas Mawson led two summer voyages in 1929–31. His exploration became the basis for a claim to sovereignty over 42 per cent of Antarctica for Great Britain under the control of the Commonwealth of Australia. Some claims were disputed: Argentina claimed the Falkland Islands; Chile and Argentina claimed overlapping sectors of Antarctica, and their claims overlapped the British sector. New Zealand also made a claim, and the sector between 90°W and 150°W remained unclaimed.

A few nations began to expand their research in the Antarctic. Australia set up the first ANARE (Australian National Antarctic Research Expeditions) base at Atlas Cove on Heard Island in 1947. Australia's first mainland base came in 1954 when facilities from Heard Island were shifted to Mawson in MacRobertson Land, the first major scientific base on the continent.

Through the period of the Cold War, tensions between the superpowers led to some uneasiness as to how Antarctica might be utilized in a time of conflict. Despite that pressure, the International Geophysical Year (IGY) went ahead in 1957–58, and in many ways it heralded the turning point for Antarctica. This was a program of co-operative scientific research in Antarctica between scientists from 67 countries. The IGY led to two important things. It led to the formation, under the auspices of the International Council of Scientific Unions (ICSU), of the international Scientific Committee for Antarctic Research (SCAR). The level of scientific co-operation during the IGY was such that it quelled political tensions in Antarctica. Governments then took the opportunity to foster harmony and subdue any territorial claims that expansion of the scientific bases might have led to. Representatives of twelve nations met for negotiations

Frank Wild (left), Sir Ernest Shackleton, Douglas Mawson and John King Davis in London in 1914. All were part of the British Antarctic Expedition led by Shackleton in 1907–09.

Chance meeting: the West German ice-strengthened resupply vessel *Icebird*, on charter to the Australian Antarctic Division and en route to an official inspection of the Soviet base Mirny (which is located in Australian Antarctic Territory), salutes a Soviet ship leaving the base.

An Australian and a Chinese scientist work together sampling water from below the sea-ice near Australia's Davis base.

that ultimately became the basis for the Antarctic Treaty which was signed 1 December 1959; it came into effect in 1961.

The original signatories to the Antarctic Treaty were Argentina, Australia, Belgium, Chile, France, Japan, New Zealand, Norway, South Africa, the Union of Soviet Socialist Republics, the United Kingdom and the United States of America. This group together with another eight nations are the so-called Consultative Parties, all involved in substantial scientific research activity. There are also now another eighteen Non-Consultative Parties which have no voting rights, are not involved in substantial research, but nevertheless are party to the Treaty. It is worthy of note that Treaty membership doubled between 1981 and 1984, numbers presumably reflecting the growing interest in Antarctica, particularly in Antarctica's resource wealth.

In many ways, the Treaty has been an outstanding international document and has, by some counts, been successful. The aim of the Treaty was clear: to ensure that Antarctica would only ever be used for peaceful purposes. It prohibits military activities, nuclear explosions and

Map of Antarctic bases

the dumping of nuclear waste. It allows for freedom of scientific investigation and sharing of observations and results. There is provision for the Treaty to be reviewed in 1991 if a member state so requests.

Importantly, Article IV, the cornerstone of the Treaty, addresses the question of territorial sovereignty only to the extent that it does not recognize, dispute or establish territorial claims. In other words, it puts aside conflicting claims while maintaining freedom of access for scientific purposes by all nations party to the Treaty. The establishment of scientific bases thus cannot strengthen a nation's claim to territory.

In recent years the Antarctic Treaty has been challenged through the United Nations. Indeed the notion of control of Antarctica by the League

Law base in the Larsemann Hills, an Australian summer research station, was opened in 1987. Such bases utilize portable fibreglass 'apple' shelters. Law base will support geological and biological studies of the area.

An Adélie penguin observes scientists near Davis base.

of Nations was mooted by Shackleton. The developing nations have for a long time expressed interest in Antarctica's resource potential. In 1983 Malaysia took the lead by having Antarctica placed on the agenda for the 38th session of the General Assembly. The issues discussed subsequently included the emerging concept of the common heritage of mankind and the suggestion that Antarctica's resources be assessed in the same way as the resources of the seabed and outer space. The Treaty nations have made some concessions to the Non-Aligned Movement of nations outside the Treaty, but in general they have been able to block such moves or have been non-co-operative by refusing to vote on the matter in the UN General Assembly. While the notion of 'common heritage' is being supported by the developing nations, the Treaty nations reject it — not through public-spiritedness or real concern about the future of Antarctica as a region of great beauty or global scientific importance, but rather so as to ensure for themselves the major share in the wealth that exploitation of Antarctica's resources could bring.

The Antarctic Treaty has always been a limited document, though it has been successful in promoting international scientific co-operation and data exchange and in 'freezing' disputes over territorial claims.

Swedish Hagglunds oversnow transports near Mawson base

However, it does not adequately deal with many emerging issues. There is no comprehensive conservation strategy for the continent, quite an omission since there is now a convention that deals with the potential for mineral resource activity. There are no common environmental guidelines for the management of bases in the Antarctic, nor for the number or types of airfields that might be constructed. The Treaty has no jurisdiction over blatant mineral exploitation under the 'banner' of scientific research (similar to the manner in which the Japanese still hunt whales) and the increasingly obvious lack of controls over independent tourist ventures and bases has not been dealt with by the Treaty nations in the twenty-eight years since its inception.

TOURISM

As man's capacity to travel around the globe and the desire to visit and explore new destinations increase, pressure to open Antarctica to tourism will inevitably mount. Certainly Antarctica is a magnificent place, a vast wilderness, arguably pristine, apart from the few scientific bases that have been established in relatively recent years. All of the sought-after characteristics of a tourist destination are represented, particularly for those people who like to choose out-of-the-way places and who are willing to forgo a few creature comforts: they can find isolation, spectacular scenery at almost every turn, and a plethora of wild animals with seemingly no fear of man.

The only things missing — at this stage at least — are high-standard accommodation and other facilities which we normally associate with tourist resorts, and of course transport and access facilities which, while available for scientific purposes, have not been developed for commercial enterprises.

There have been some tourist ventures into the Antarctic since the late 1950s. Aircraft were occasionally chartered from Chile and Argentina to overfly the Antarctic Peninsula. In the late 1970s, large commercial jet aircraft from Australia and New Zealand made forty-five flights southward enabling thousands to see the Ross Sea region of Antarctica. In November 1979, 257 people aboard an Air New Zealand DC10 were killed when the plane crashed into Mt Erebus. This tragedy ended large-scale tourist flights over Antarctica.

Most visitors to Antarctica arrive by ship, and most are scientists or service and support staff for the established bases. The United States and New Zealand, however, utilize air transport provided by the USAF, particularly to the McMurdo and Scott bases respectively, near the Ross Sea. Argentina, Chile and the Soviet Union also have the capacity to fly personnel into the Antarctic region during the spring and summer.

In the 1960s it was anticipated that most visitors would be participants in government programs or at least on expeditions with government endorsement. Nowadays private expeditions using large yachts and

Lindblad Explorer

chartered ships for transport occur regularly, and tourist visits have increased in number, particularly in the last ten years. Cruise liners have now taken many thousands of tourists from South America into the region of the Antarctic Peninsula, mostly using the *Lindblad Explorer* as a platform for sightseers.

International conservation groups such as Greenpeace also visit Antarctica. Greenpeace has established a 'World Park' base and routinely uses its ships to service the base and monitor environmental issues such as the construction of the French airstrip in Adélie Land and the continuation of whaling by the Japanese.

In recent years light aircraft have also been used by private groups to gain access to Antarctica. At the Chilean station at Teniente Marsh there is an airstrip serviced by a tourist hotel, and it seems likely that private light aircraft flights from Australia to eastern Antarctica will be expanded in the future.

The number of tourists is not known, but most have chosen to visit Antarctica because of their fascination or reverence for the continent.

Because of their respect for environment there have been few problems. There are regular tours organized for bird observers and other groups, particularly to the Antarctic Peninsula where there are large numbers of animals. Other locations of interest are the few historic sites and other sites of special scientific interest such as penguin rookeries.

It is important to be aware that although Antarctic Treaty parties are technically responsible for their own nationals participating in private expeditions, in reality it is difficult to control such projects. Certainly for tourist ships registered by non-Treaty nations, control is nil.

Tourist ships have occasionally run aground in Antarctic waters and on a number of occasions private expeditions have come to grief and required assistance from personnel at an established research station. This assistance is always expensive and disruptive but morally these establishments are obliged to assist.

The question of access to sensitive environmental areas is also under question. There is nothing to stop a commercial tourist venture from establishing itself adjacent to a particularly vulnerable animal habitat. At present the greatest deterrent to such enterprises seems to be cost.

The uncontrolled spread of tourist activities and facilities is a problem that must be addressed. Already the establishment of scientific bases imposes significant pressure on the limited number of ice-free sites on the continent. Such locales are often those used by birds and seals as their breeding sites and as such are favoured by tourists for their visits. The impact of human foot traffic on fragile vegetation communities in these crucial areas is severe, disruption to penguin and other bird rookeries can also be significant, the impact of noise from vehicles and overflying aircraft has not been established and there is the continuing problem of waste disposal. It would be ideal if all tourist operations remained ship-based with only short-term incursions onto the continent itself. Such floating hotels would also facilitate the complete removal of rubbish and all human waste from the Antarctic zone, returning everything to its country of origin. The consideration of Antarctica for World Park status would also encourage the regulation of tourist behaviour.

Sunset over ice near Rumdoodle, Framnes Mountains

CONSERVATION

The 'Green Factor' is now recognized worldwide. There is a rapidly growing sense that man is putting the planet at risk. The issues crucial to Antarctica are echoed throughout the planet: considerations of shared resources, resource-use against the value of wilderness, the need for full-scale scientific models and base stations for global climate-modelling, and the need to retain an area free from any political conflict. Antarctica plays a valuable role in highlighting these issues but the limits to the continent's ability to absorb pressure from development by man are being tested. It is those values together with the more obvious ones of the fragility of the Antarctic ecosystem and the simple concept of retaining at least one continent in a wilderness state that have led to the proposition that Antarctica be declared a World Park. However, all nations want to share in Antarctica's wealth. The stakes are now much higher than when the explorers, and sealers and whalers sailed the Southern Ocean. Today the prizes are minerals and oil, commercial fish reserves and perhaps a strategic advantage through control over Antarctica.

The Antarctic Treaty nations have been able to keep the issue of Antarctica 'low-profile' and conceal most of their discussions by holding them in camera. Conservation groups and scientific organizations are not permitted to attend Antarctic Treaty meetings. Despite their efforts, Antarctica is now bound up in one of the greatest international power struggles the world has seen. A few small scientific stations on the continent allow Antarctic Treaty nations to raise their flags, and these may become the first few cells of a cancer that would irreparably change Antarctica.

The original Antarctic Treaty did not deal specifically with protection and conservation of Antarctica's animal and plant life or the Antarctic environment itself. However, the Treaty parties have since set up Agreed Measures for the Conservation of Flora and Fauna. There is a Convention for the Conservation of Antarctic Seals and a Convention for the Conservation of Antarctic Marine Living Resources (CCAMLR), designed to regulate commercial fishing operations in Antarctic waters.

The Greenpeace ship *Gondwana* in pack-ice

In an effort to gain further insight into the interdependence of the species in the Southern Ocean ecosystem, scientists set up BIOMASS, the Biological Investigation of Marine Antarctic Systems and Stocks. BIOMASS examined, for example, whether krill fishing was interfering with the recovery of whale populations, or whether bird numbers could be correlated with the location and size of krill swarms.

Scientists remain concerned about the success of BIOMASS and CCAMLR. It has been difficult to enforce the regulations. Despite the conventions, data on fish catches have not been provided, either for the period before CCAMLR was established, or subsequently, in time for meetings of the committee set up to consider the data.

Once again, while scientific co-operation operates at a high level, political co-operation, particularly when commercial interests are involved, leaves something to be desired.

Fin-fish and krill are still harvested at unacceptably high rates. Trawling around the sub-Antarctic islands of Kerguelen and South Georgia began in the 1960s. In the early 1970s enormous numbers were taken by East German, Polish and Soviet ships. Even in the area controlled by CCAMLR, there are no regulations regarding mesh-size and many immature fish are caught. Also, Antarctic fish species take many years to mature. Lack of regulation and the over-fishing have threatened the long-term survival of some species. It was recorded that 162 598 tonnes of Antarctic icefish were caught in 1982/83; a catch of this magnitude may have significantly threatened future stocks, but we do not have sufficient information about icefish or their total populations to understand properly the impact.

Krill harvesting by Japanese and Soviet vessels also began in the 1960s. In 1981/82 the total krill catch peaked with a haul of 528 000 tonnes. Since then totals have been lower but are still some hundreds of thousands of tonnes annually with Poland also now a major harvester. Krill is used largely for animal feed, in particular for the fur trade in northern Europe. Krill for human consumption was a potential saviour of developing countries with protein shortages in diets. An early problem of high fluoride levels in the exoskeleton of the krill has been solved by snap-freezing the catch, and now krill can be processed and canned on board ship. Technology has propelled renewed interest in krill fishing, but it has been found that the life-cycle of krill is longer and productivity is less than first thought, so now there are questions about the ability of krill swarms to re-establish themselves in commercial quantities. Now that we are learning more about krill and its role in the Antarctic food chain, it seems that large, sustainable catches will not be possible in the future without serious disturbances to the Southern Ocean ecosystem. It is hoped that the large catches already made have not significantly altered the balance of nature in the region.

Soviet and Japanese whaling fleets continued to operate through the 1970s and early 1980s. An International Whaling Commission moratorium on whaling came into force in 1986; however, the Japanese still take hundreds of minke whales and some sperm whales for so-called scientific purposes.

The limitations of CCAMLR and the apparent disregard for the other international conventions that has been demonstrated by over-fishing and 'scientific whaling' have created concern among many environment, conservation and scientific groups and organizations about the potential success of any new Antarctic convention.

CRAMRA, the Convention on the Regulation of Antarctic Mineral Resource Activity, governs the possibility of minerals development in Antarctica. In the early 1970s, interest in prospecting in Antarctica was

Lesser sheathbill, endemic to Heard Island

expressed by marine geophysical companies and the issue was brought up at a Treaty meeting in 1972. Then in 1973 the *Glomar Challenger* discovered traces of methane and ethane (often associated with the presence of oil resources) in the Ross Sea. At the same time OPEC established an oil embargo that caused a significant increase in world oil prices, leading to greater interest in developing new oil sites.

After six years of negotiations, CRAMRA was adopted on 2 June 1988 and was open for signing and ratification by the Antarctic Treaty nations from 25 November 1988. Many nations promptly signed the document. To be ratified it had to be signed within twelve months by 16 of the 20 Consultative Parties to the Treaty, including 11 developed and 5 developing nations. Signatories had to include the two superpowers, USA and USSR, all of the seven nations that claim Antarctic territory, and at least 7 additional non-claimant nations of which 3 must be developing nations.

As an original signatory to the Antarctic Treaty and as one of the seven nations that claim Antarctic territory, Australia was required to sign the CRAMRA agreement before it could be ratified. The Australian Conservation Foundation as well as international conservation groups such as Greenpeace led campaigns to convince the Australian government not to sign. French naturalist Jacques Cousteau on behalf of the 300 000-member Cousteau Society and Foundation Cousteau also appealed to the Australian prime minister to set a moral example by refusing to sign. A similar appeal was made by Sir Peter Scott, son of British Antarctic explorer Robert Falcon Scott.

As yet no commercially exploitable mineral deposits are known to exist in Antarctica and there are no known plans to begin exploration for them. The new convention is to be in place prior to such discoveries. It is not designed to encourage minerals activity, but rather to set out a regime that would avoid conflict of interest as well as safeguarding the Antarctic environment if and when minerals activity begins.

There are, however, large deposits of iron-ore in the Prince Charles Mountains in the Australian Antarctic Territory, and perhaps the world's largest deposit of coal in the Transantarctic Mountains. There is considerable interest in the Dufek Massif which is similar in geological structure to the Bushveld complex in South Africa, a very rich mineral site. There are also traces of gold, silver, titanium, tin, copper, cobalt, molybdenum, manganese, nickel, platinum, lead, zinc, and uranium. At this stage very few of Antarctica's ice-free areas have been geologically mapped and knowledge is limited. Oil reserves are possible, but not confirmed.

The problems involved in extracting and marketing any minerals from the Antarctic environment would be enormous. Weather conditions would be the first obstacle; then there would be the problems of icebergs and water depth for offshore oil wells, extremely thick ice and access for onshore sites, and storage and transportation of the products. Many people believe that extraction of Antarctic minerals would be totally uneconomical and that exploration and development are therefore very low priorities for the mining companies.

Scientists tag elephant seals with plastic cattle tags in their back flippers to determine migration patterns and population numbers.

On the other hand, international environment organizations suggest that strategic and political motivation could outweigh economic considerations. They point out two important provisions in the CRAMRA agreement. The first allows for limited liability for environmental damage should, for example, an oil spill occur. The second allows for confidentiality of data regarding prospecting and exploration.

More importantly, there is the notion that simply by its existence the CRAMRA document focuses attention on the possibility of mineral wealth in the Antarctic, thus encouraging activity. The possibility of oil pollution, either from offshore wells or leaks from tankers, would be devastating in the Antarctic environment. Little is known about the toxicity of oil to Antarctic species or the potential duration of impact. Estimates put the long-term effects of an oil spill at more than a hundred years.

The March 1989 oil disaster in Alaska must hold some lessons for those wishing to develop the Antarctic. When the oil tanker *Exxon Valdez* ruptured after hitting a reef in Prince William Sound, 1.2 million barrels of crude oil were released into the water, devastating the Alaskan coastline and endangering whales, seals, sea otters and both migratory and resident sea-birds. Described as the worst oil disaster in North American history, it demonstrates that despite technological safeguards and agreements, environmental disasters still occur.

Following the accident, the French prime minister declared that France would not sign the CRAMRA agreement because the *Exxon Valdez* disaster exposed significant flaws in the convention.

Any development in Antarctica, for exploration or exploitation, will require much larger service facilities than are currently the norm there. Such development would put increased pressure on the few ice-free areas along the coast and would be further competition for the birds and seals that use them as breeding sites.

The Antarctic and Southern Ocean Coalition (ASOC), which includes Greenpeace and the Australian Conservation Foundation, continues to propose making Antarctica a World Park as the best option for the future of the continent. Their concept is based on four main principles: firstly that the wilderness values of Antarctica should be protected; secondly that there should be complete protection for Antarctic wildlife while allowing limited fishing; thirdly that Antarctica should remain a zone of limited scientific activity, with co-operation and co-ordination between scientists of all nations; and finally that Antarctica should remain a zone of peace, free from all weapons.

Greenpeace World Park summer station

Mawson, the first permanent station in Antarctica, was established in 1954. *Icebird* lies in Horseshoe harbour while resupplying the base; the Antarctic icecap and the Framnes Mountains are in the distance.

World Park, World Heritage, world resource?

ACKNOWLEDGEMENTS

This book represents a condensation of thoughts and ideas that evolved subsequent to my visit to Antarctica. On that visit, and in the ensuing period, I developed a considerable interest and empathy for Antarctica as it was and as it might be.

I had wanted to visit Antarctica for many years and I am therefore indebted to the Channel 10 network for the opportunity to represent the network when the offer to travel to Antarctica with the Australian Antarctic Division was made. To cameraman Chris Brown and his assistant Steve Welch who travelled with me, I am indebted for not only companionship, but for the moving pictures that made the Eyewitness News special 'Beyond the Frozen Edge' and the subsequent television documentary 'Antarctica: Who'll Break the Ice?' so successful.

To the members of the Australian Antarctic Division, in particular those who travelled on Voyage 6 1987, go my sincere thanks; I especially wish to thank Peter Keage and the late Dr Gavin Johnstone for their assistance in the development of my appreciation for Antarctica. To the others who travelled on Voyage 6, and to Captain Ewald Brune and the crew of *Icebird*, I also extend my thanks for what was a most enjoyable voyage.

My interest in and knowledge of Antarctic matters were fostered and maintained by a number of people, but in particular I must thank Margaret Moore, the former campaign officer of the Antarctic Action Group of the Australian Conservation Foundation.

Ross Scott provided me with the photographs of the lichen in Lichen Valley, Vestfold Hills; the leopard seal at Heard Island and the emperor penguin. Dr Steve Nicol of the Australian Antarctic Division kindly made available slides of Antarctic krill. Thanks to Lyn Goldsworthy and Greenpeace for the photograph of the *Gondwana* and the Greenpeace summer base. Lindblad Travel kindly provided the picture of the *Lindblad Explorer*. For the provision of the historical photographs on page 73 and 74 I acknowledge the assistance of Dr Fred Jacka and the Mawson Institute for Antarctic Research, and for the photograph on page 77 I thank Bob Reeves of the Australian Antarctic Division.